Common Core Mathematics: Teaching Kids Common Core

Our Children's Success with Common Core Teachings

By: Julie Thompson

9781634289870

PUBLISHERS NOTES

Disclaimer – Speedy Publishing LLC

Speedy Publishing LLC

40 E Main Street, Newark, Delaware, 19711

Contact Us: 1-888-248-4521

Website: http://www.speedypublishing.co

REPRINTED Paperback Edition: ISBN: 9781634289870

Manufactured in the United States of America

DEDICATION

I dedicate this to all parents. May this book be a good guide for you to better understand the education system that may hopefully lead to the success of your children.

TABLE OF CONTENTS

CHAPTER 1- INTRODUCTION TO COMMON CORE MATHEMATICS

There is plenty of buzz out there about common core and what it means for children in public schools. Some parents are very impressed with it and love the idea. Others aren't sure about it because they have heard complaints and that it confuses kids.

Some educators really embrace it and others don't like the idea of change. They also worry that it means they have to change their teaching methods. Yet there are plenty of benefits involved with the common core curriculum. As you learn about it, you will be more comfortable with what your child will be learning about.

In the past, a big problem with schools is that they all did things differently. A child going to school in one region would learn different things than others. Yet they all took the same standardized tests. A student that moved due to family circumstances may find they are in a school where they are very far behind or where they are too far ahead.

Common Core Mathematics

With educational standards in place, educators can make sure that students are exposed to the skills and the information they need to build on from one grade to the next. Parents can feel good knowing that their child is going to get a good education no matter where they have and which public school they attend.

Educational standards help to make sure that all students get the foundation they need for a successful future. It doesn't matter what their goals and plans are for after graduation. Ensuring that children have skills they can use in the real world as well as in their post-secondary education is very important.

The quality of education is important, and the common core curriculum helps to ensure that students are able to consistently get that. The standards can be implemented from one school to the next as well as from one state to the next. It gives children a fair opportunity for learning in the public education system.

Educators and supporters believe that with common standards in place, they are better able to serve the needs of all students. Educators still have the freedom to use their own teaching style and methods within a classroom setting. However, they will have help with creating the program that they will teach.

Students in any given grade will have certain knowledge and skills that they should learn during the course of that given school year. Educators have the chance to create great lessons as well as a terrific learning environment. We all know that in any school there seems to be those teachers that do a great job. You also have those that don't teach enough and those that will overload students.

The goal of the common core is to help create a specific set of objectives and goals that educators use in a classroom to help students learn and be ready for the next grade level. With clear standards in place, parents are also able to understand what their children are learning and why in any particular grade. This enables parents to be more supportive and to assist with making school a pleasant experience for their children.

With standards in place, children can get a quality education in any public school they attend. It doesn't matter if it is a wealthy upper class neighborhood or a low income area where they attend class. It levels the playing field for those students all to build a successful future upon.

Eventually, children grow up to be adults. It is believed that the common core system is designed to help them do well in the workforce as well as in college. It is very sad when you realize that there are high school graduates out there right now that can barely read, write, spell, or do simple math. They need those skills in every aspect of their life so it holds them back.

With the standards in place, it is believed that children won't be falling through the cracks based on the teacher they had or the school they attended. This also reduces pressure on parents to live in particular areas so that they are zoned for their children to attend a certain public school with a reputation of a quality curriculum.

If you are uneasy about what the common core offers, you don't need to be. There is plenty of quality information here for you to learn from. It will put your mind to ease about what kids will be learning. It also helps you to see the value for teachers who are often overworked and underpaid.

The common core is designed to allow kids to be creative, to allow teachers to use the methods they are comfortable with, and for students to grow and to learn new concepts that will help them in the future. It isn't a punishment and it isn't going to ruin public schools. Instead, it offers better quality, detailed learning objectives, and a real chance for all students to get the same level of education from public schools.

CHAPTER 2- UNDERSTANDING OF COMMON CORE

If you are like many educators, parents, and the general public, you have heard the term common core quite often lately. People may be upset that they don't know what it is or how it affects them. Others are excited about it as they know the value it holds.

The underlying factor that makes common core so exciting is that it offers the same standards across the board. The 7th graders going to school in California will be learning the same principles and curriculum as the 7th graders in New York!

Many worry that this will lead to the Federal government being in charge of the education system. However, that isn't the goal at all. There aren't steps being taken to nationalize the education system in public schools.

Parents, educators, and the general public have been involved in developing the common core curriculum, not the Federal government. Each state will get to decide if they are going to adopt the standards or not. Almost all of them are seeing it as a favorable way to make sure students learn what they need to in order to do well in society after graduation.

They also believe that it will make the school learning environment more appealing. As a result, it could reduce behavior problems, students that are bored, and it could reduce dropout rates which are all very encouraging outcomes if the core curriculum does well.

The standards that are part of the common core curriculum are well designed. They are created in order to prepare all students for success as they advance from one grade level to the next. The program is designed to include real life skills that are at the foundation of performing tasks including work as an adult.

The core curriculum is also going to help even the playing field when it comes to the application process for colleges. Students have the same opportunities to do well in college too because they have been exposed to the same learning information during their time in the public school education system.

This is important because some people worry that the common core is going to reduce the quality of what students learn in public schools. The opposite though is what is really going to take place. The best practices will be kept while what hasn't been successful will be removed. Those elements will be replaced with better learning options and opportunities.

Part of the agreements in place for accepting the common core is that no state involved that accepts it will lower their standards. This means that the best practices from public schools around the United States will be brought together.

With so many college students right now struggling to get through the programs, it does indicate that there is a huge gap between what is being taught in many public schools and what students need to know to do well in college. It is hopeful that the common core will help to bridge students into college instead of them being in unfamiliar territory.

There is no mandatory acceptance of the common core curriculum. The fact that almost all states have done so though is very encouraging. It helps to provide a united effort for admin, educators, parents, and teachers.

Those that oppose the common core though feel that it doesn't prepare students for an upscale type of education at a four year college. They feel it won't prepare them for the STEM careers. This can be worrisome for some parents that feel that their kid may not get into a good college program because they didn't learn calculus in high school.

Keep in mind that all schools still have the option of offering advanced classes for students that test very well within the core curriculum. The program isn't going to hold them back and allow them to get bored. For example, advanced students in math can go to a separate class and learn advanced math while other students learn what is in the common core.

The International expectations of students are also a part of the common core program. Many experts have long stated that our students in the USA fall far behind students in some other countries. With the global use of the World Wide Web for business and education, students need to leave high school with the ability to compete on an International level. It is hopeful that the common core will give them a firm foundation for that to occur.

Common Core Mathematics

Some will argue that the common core doesn't offer enough, but others will argue that it is a good starting point. It should be viewed as the building blocks for the future of public education to continue with moving it the right direction. Changes can't occur overnight, but this is definitely doing more than just spinning wheels or pointing figures. That type of behavior has gone on far too long in the public school systems.

CHAPTER 3- COMMON CORE - CHANGES IN THE CLASSROOM

In order for the common core to really work, everyone has to be prepared for it. A positive attitude and knowing what it does and doesn't offer is important. Many public schools are holding meetings and forums that offer this information. They invite parents and the community to come learn what it is all about.

Administrators

Administrators need to guide the path for their educators when it comes to the common core. They need to work with teachers to make sure they have the tools and resources in the classroom to teach what the common core indicates that they need to. They also need to reassure educators that they will have help should they need it.

Administrators need to ensure that they don't have staffs that aren't willing to embrace it. The common core program will only work if everyone is on board. If you have a handful of teachers that resist it, then problems are going to develop.

Educators

As more educators take to the common core curriculum, they see the value it offers. They feel it helps them to teach as a unit rather than as an individual. They can also eliminate the frustration of students coming to them from the previous grade and not being ready to learn what they are going to teach.

Educators find that the common core helps to increase the standards of what students are learning. Teachers certainly aren't in this career for the money. They have a sincere desire to help students to do their very best. They will find that students have the building blocks they need to be successful and to be confident as they move into the next grade.

Some educators though feel that the common core curriculum steps on their toes. They worry that their freedom to be creative and to teach in a manner that fits their personality is being taken away from them. This isn't true at all though. They can still teach the way they would like to but they do need to make sure they are covering all of the standards that the common core includes.

Parents

Many parents find what their kids are learning is over their heads. They can't help with math homework because they don't remember how to do those formulas or equations. The common core can help parents to get back into the loop. They will know what to expect with each grade level.

That is important as too many parents feel that they can't do enough to help their kids get a wonderful education. Parents that don't live in regions with the best public schools often feel that they have failed their kids. They are angry and disappointed that they aren't going to get the same level of education.

Now they can experience peace of mind that it will no longer be a barrier in the path of the future for their child. With the 15% above level standards for advanced students, they also don't have to worry that a child with advanced skills in given subjects will be held behind to stay with everyone else.

Even those schools that don't have enough teachers to offer a higher English or math class can offer solutions. For example, online learning with a computer or an iPad. With the standards being the same, schools can also pool resources so that it is possible to get these advanced students what they need without budget restraints.

Students

School can create anxiety for students that feel like they are behind. The common core helps to eliminate that should a student change schools. It won't matter if they move across town or across the USA. They will be able to stay current with the curriculum for that grade level.

Students will also enjoy the fact that they get the same opportunities for a quality education no matter where they live. Being well prepared for college and for the workforce are very important things. Very young students may not be thinking about that future yet but it will be there before they know it.

Common Core Mathematics

Students that go out into the real world lacking skills may struggle to make ends meet. They may find that they don't do well in college and drop out. That can really hinder their self-esteem. It can also put them on a path of dead end jobs and living from paycheck to paycheck.

When students enjoy school, they will do better. They will pay attention and they will be willing to learn. They will also do better on standardized tests which is important. It shows that they are learning, that they have made growth within a calendar school year, and it will help them to create the future they want.

CHAPTER 4- COMMON CORE IS POPULAR NOW IN ALL DIFFERENT SCHOOLS

This is why the common core is so popular right now. Standards are being developed so that students can do well in college as well as in their careers. It is anticipated such changes in the public school system will help to improve the contribution that future generations can make to society as a whole.

Both educators and parents will know what they can do in order to help students learn. They will also be on the same page when it comes to what students should be learning. Parents can also help prepare their children for the upcoming school year by working with them on the known concepts for the next grade during summer break.

With the consistency in place, students no longer suffer from a poor education due to where they live. Since no state will lower their standards, students that live in the regions will better programs will still continue to get a quality education. However, it will also be spreading around the USA too and that will increase the learning benefits to all students out there nationwide.

The curriculum includes content as well as application. This allows the knowledge that students gain to be applied with high order skills in place. It allows them to build upon a foundation that was created in the lower grades of their public school education.

Common core is both evidence and performance based. This isn't just another idea to toss around. It is a tried and true method that has been constructed from evaluating the benefits of great public school programs and also the pitfalls of those that had plenty of shortcomings.

While common core is USA based, it is believed that it can be a global benefit as well. Students with a solid foundation of learning are able to use those skills and abilities to help on a global scale. The volume of technology in place today means that global business and worldwide economy are very intertwined. Much more so than they have ever been in the past. That growth is projected to continue into the future too.

It is believed that the common core is a very realistic way for public education schools to operate. It has already been quite effective in the classroom based on the pilot programs. Not only were students feeling motivated to learn, but educators reported feeling less stress about coming up with lesson plans.

Students in these pilot programs also did very well on standardized tests. This was conducted to see what the level of learning was during the timeframe of the pilot program. The students were given the same test at the start of the pilot program as they were at the end of it. This allowed the test results to show the gains of each student during that period of time.

Julie Thompson

The common core public school curriculum has been accepted by the following:

- 4 Territories

- 45 States

- Department of Defense Education Activity

- District of Columbia

Common Core Mathematics

The 5 states that have not yet decided to accept the Common Core State Standards are:

• Alaska

• Minnesota

• Nebraska

• Texas

• Virginia

Many people in society feel that students don't have an equal chance when it comes to being ready for the workforce or for college. The design of the common core is to make sure those students all have the same background and foundation to work from. It doesn't matter where they lived or where they went to school. The program is also going to make it easier for children that move due to their family situation, changes in family work, or even military families to be able to stay in sync with their education.

The biggest areas that will be aligned within the common core are math and English. The standards that are in place for them helps to keep students on track for each grade level. This is important because before you had some students learning algebra in middle school and for others it wasn't until high school.

The same was true with the English standards. In some middle school curriculums, students were reading classics and writing papers with citations and resources. In other programs, such elements of the curriculum weren't introduced until the last couple of years of high school.

So why have some states selected NOT to be part of the common core curriculum? There is plenty of speculation about that! However, what is known is that most of them didn't accept it. For example, in Texas, the vote was 140 for it and 2 against it. They don't go with the majority rules for the Texas House of Representatives.

CHAPTER 5- COMMON CORE MATHEMATICS AND ITS STANDARDS

Creating math standards is one of the most important parts of what the common core is all about. It offers a better overall focus and coherency between one grade to the next. Children in the early grades will be taught about whole numbers, operations, and relations. They will also learn the basics of shapes, measurements, and geometric.

However, more of the learning is going to be contributed to focusing on the actual numbers involved in math. Think about all of the math entities you took part in when you were in school. Yet it is really all about the basic numbers for most people in their daily tasks and in their jobs. There are exceptions such as those that are in a career that involves high levels of math and high levels of science.

The K-6 grade math standards will help with creating a solid foundation for them to expand upon in middle school and high school. The goal is for approximately half of the time to be dedicated to working directly with numbers. There will be many ways in which this takes place in it is completed.

The problem with many of the math programs before common core is that they are weak. There are too many mechanics involved and that can make it tough on students. Not everyone learns the same way, and the common core allows for several ways for the same math equation to be done. This reduces the chances of kids getting frustrated or getting behind with the learning concepts.

The math curriculum can be organized in a variety of ways, depending on the grade and the teacher. Too often, the way that math is presented now is much distorted. As a result, students feel intimidated by it. This can give them a negative attitude towards math concepts in the future.

Research that has occurred in the past 10 years shows that the area of math is very below average in the United States for public schools. With the common core, the standards are going to address such problems. It is going to be a challenge, but it is going to help change the way that students learn math as well as what concepts they are exposed to.

The fact that all students in a given grade will be learning the same math concepts is an important part of such changes. It helps ensure that students have the right information before they enter the workforce as an adult or before they go into a college math class.

The standards are going to offer clarity and they are going to be very specific with the common core. The standards that will be applied are going to make assessing math skills for public education students a simple process in the future. At the center of the standards will be the principles and the values of the laws of arithmetic. The rest of the curriculum will be structured as branches from that.

Many students are currently challenged by math concepts that they just don't understand. They aren't getting the meaning that is underlying for them to study and to fully develop those skills. The standards are researched based and the learning will progress for public school students in the common core from one grade to the next.

The result is going to be for students to learn and to be able to retain what they learn in math class. They will have knowledge and skill as well as critical thinking abilities. They will also be able to take those skills and concepts and see them grow with each additional grade that they complete.

Educators, parents, and students will all be able to define what should be learned in a particular grade. This justifies changing the curriculum so that everyone is on the same page with all of it. This means that students aren't going to be shortchanged with the math skills due to a teacher not sharing with them certain elements of the program or that they live in an area that isn't offering the higher standards in math.

With the common core, students will be able to understand the math procedures. There will still be programs and resources in place to help any students that are behind in the math curriculum. That way they will be able to get caught up and move forward with it into the next higher grade level.

Special needs students will also need to be assessed differently so that they can learn math in a manner that works for them. There are several assistance types of devices that can be programmed for the common core of math for any particular grade level. Individual aids and tutors can also help with learning math and staying on track.

As previously mentioned, students that seem to be advanced in their math skills can be placed into higher level learning environments for math. The guidelines of the common core though state that it can't be more than 15% higher than the standards in place for that particular grade level.

Most educators though agree that the 15% upgraded learning option for math is more than adequate per grade level. They feel it gives them the room to ensure that all students learn math and that those that have advanced skills have a way to continue being challenged. It is a winning outcome for all.

Chapter 6- Common Core English and Its Standards

The other major area of learning within the common core curriculum is in regards to English standards. Technically, this is in reference to English Language Arts & Literacy. The standards are the collective efforts of an extended effort to create very high learning in this area for students in the K-12 public school systems.

Literacy is a big part of school and students that struggle with it often don't graduate from high school. They are easily frustrated, lack self-esteem, and they struggle in all of their classes due to the volume of reading that has to be done. A strong foundation in literacy and English from the start can result in learning being fun and not a struggle.

Too many high school graduates struggle with literacy when they enter the workforce. They may get a shock when they see the volumes of what they have to read and all they have to write. They need to have skills that allow them to write reports and to complete all of the paperwork that is part of their job.

Too often, high school students struggle with literacy and English classes at the college level too. If they weren't involved in a program that had high standards then they may fall short that very first semester of college which can be a huge blow to their future plans.

The curriculum standard for English and literacy within the common core is under the leadership of the NGA (National Governor's Association) and CCSSO (Council of Chief State School Officers). The best models are being used and so are research gathered from:

• Assessments

• Educators

• Parents

• Professional Organizations

• Public Citizens

• Scholars

• State Departments of Education

• Students

The result is a standard that is very well rounded based on such feedback and direction. There are several standards in place for the

Common Core Mathematics

English and literacy common core curriculum that the NGA and CCSSO have accepted. They include:

• Standards are evidence and research based

• Standards are parallel to what students will need in the workforce and in college

• Standards are building blocks and advance with each grade level

• Standards have specific benchmarks of learning that need to be met

The standards will be identified for each grade with specific milestones being met. The standards will cover reading, writing, speaking, and listening skills. Improving literacy in the areas of history, science, and social studies is also part of this literacy plan too.

The literacy plan is created to help ensure that every student that completes the 12th grade has the skills they need for workforce and college readiness. Being able to take those skills learned and apply them outside of a classroom setting is a goal of the common core.

Selecting literature that is high quality in the classrooms of public schools is being closely evaluated. Not only should it help with reading and with English skills, it should also help with broadening the concepts and thoughts about the world in the eyes of students.

Another goal is to promote creativity with effective reading and writing skills. Students should be able to use the skills that they have learned and to apply them to create poems, short stories, and essays that are appropriately assigned for their grade level.

What is disappointing through with the common core curriculum relating to English and literature is where the trends will be going. The reading time of classic literature will be reduced by 50% for high school students. That reading will be replaced with informational types of text materials.

There is no set reading list for each grade level through the common core. Teachers and admin will need to make sure what is selected to be read and discussed in a given grade level is going to be appropriate. This can be an area where there is plenty of gray rather than it being in black and white. Some believe this is an area of the common core that will need to be improved upon in the future to prevent debates and issues about reading materials.

However, the changes are believed to be a way for International benchmarks to be reached. It is no secret that there is such a big gap between what USA students offer globally compared to some other countries. Being able to close that gap so that they can successfully compete on a global scale is very important for their futures.

The value of content knowledge isn't lost with the common core English curriculum. That is a myth that seems to continue to be circulating. The goal is to teach students how to analyze so that they can take the skills they learn and apply it to real situations out there in the world.

Students will find the English and literacy areas of learning to be more interactive and hands on. It isn't going to be boring and just sitting there. For kids that aren't even reading, just spacing off for the time allotted, then this can really challenge them to stay on task and encourage them to learn.

Chapter 7- Common Core and the Parent-Institution-Relationship

It makes sense that parents have a vested interest in what their children are learning. Working with the school to make sure that the common core curriculum is in place and supported is important. The admin and the educators need all the help they can get. Don't assume that they do all the teaching and you do all the parenting.

Work with them to make sure you understand what will be taught in each subject for the grade each of your children is in. This will allow you to prepare at home too with learning objectives and even when you need to lend a hand regarding their homework efforts.

Ask questions about the common core if you aren't sure about things. Don't make assumptions and don't listen to rumors or myths. Just because you find certain information online, that doesn't make it the truth. Get involved and find out what the program means for your family and for the students of the school as a whole.

Give your Comments and Feedback

The school educators and admin know what you like and what you don't about the common core. With that input, things that do work can be solidified. Those that don't can be modified. Both NGA and CCSSO have public comment periods so that parents can offer comments and feedback to them.

Ask About Teaching Methods

As previously mentioned, the common core doesn't direct how teachers implement the standards in the classroom. As a parent, don't be shy when it comes to asking about teaching methods that will be used. If you find that your child isn't doing well with the methods used, come up with a plan that enables your child to learn in a manner that does work for them.

If the teacher isn't open to that, talk to admin. It may be a decision that requires the child to be moved into another classroom. It could be a discussion that leads admin to request that the teacher modifies their teaching techniques. Especially if they are hearing similar concerns from more than one parent.

Ask teachers about their lesson plans and what they plan to do that will help the curriculum to work. Many schools will hold forums where the admin will share what they hope to accomplish in the school with the common core curriculum.

Common Core Mathematics
Find out the Standards for their Grade

Most schools offer a back to school night where you are able to meet teachers. This is usually in the first couple of weeks of the school year. It is the ideal time to find out the standards for that grade. You should also be able to find that information online. However, the teachers at your child's school will be able to explain to you exactly how they plan to teach the students to meet those standards.

Let them Know you realize there are Changes

Parents have to be realistic about how common core will change things. Kids aren't going to immediately do better and have better test scores. It is going to be a process of changes over time. Be encouraging both to your school educators and to your children. Talk positively about the common core so that kids will see it as an opportunity for them.

Tests may seem harder at first for kids due to the changes. This is because more may be expected with some of the schools that were performing lower than the standards before. It is going to take more time for them to get caught up and on level than it will for the students in schools that were already performing at the standards or very close to them.

The Smarter Balanced testing exams will begin to be allocated for the 2014 – 2015 school year. Students need to relax and not worry too much about the testing process. Instead, they need to get plenty of rest, eat breakfast, and be encouraged to do their very best. Parents should do what they can to reduce activities and outings during the test week to make it low key for students.

Not Federally Promoted or Funded

It is important to understand that the funding your child's school gets from the government isn't based upon their decision to go with the common core or not. However, there have been some grants offered to schools for programs and resources that they would like to have in order to successfully implement the common core curriculum.

Some schools have applied for grants but also asked parents to help with fundraising in order to get what they need. The school doesn't want the children to be without what they need due to a lack of funding. A cut back in the schools has been a serious problem for quite some time.

CHAPTER 8- CHILD'S SUCCESS WITH COMMON CORE

With all schools that fall under the umbrella of the common core on the same page, it makes evaluating the success of a child much easier. They can all be evaluated using the same set of standards that will be put into place.

Standardized Testing

All public schools currently use standardized tests to gauge where their students are at. Should a school have low scores, they often get reprimanded by their state department of education. Since there are so many different curriculums, it definitely makes it hard to evaluate correctly though.

For example, students may be learning, but not learning what is on that test. That doesn't mean they didn't learn. It just means they didn't learn the right things that the state is testing them for. It is like comparing apples to oranges and expecting to come up with something that is the same. It simply won't work!

With the common core, there will be standards in place for learning of each subject area for each grade level. Those learning objectives will be consistent with what is on the standardized tests. So it is comparing apples to apples, and students either know the information or do well on the tests or they don't. If they don't, then the state department of education can contact that school and work with them to revamp what they are teaching.

Admin & Educators

Happy staff is going to be a great way to evaluate the quality of the common core. Many teachers complain that they work too hard and too many hours. They are stressed about standardized tests and about coming up with lesson plans. With the common core, they know what they have to teach and there isn't any guessing involved.

Many educators can feel relaxed and they can plan their lessons accordingly based on those standards that are in place. Admin won't be looking over their shoulder as much to make sure they are teaching what they should be. This makes for happier teachers and to also free up time of admin to focus on other tasks.

School Environment

The overall school environment can improve with the common core implemented. The school may have a terrible reputation now due to the programs they teach or the methods. It is time for a clean slate to get things moving forward. With common goals of everyone, there isn't the push and pull tug of war taking place.

Instead, it is a collective effort for all to engage in. Arguments about what to teach will no longer be an issue. There will be a checklist of what should be covered in each subject in each grade.

There won't be any more comparing this school against that school. There won't be any more issues with students that are behind when they move to another school district.

Student Happiness

Students that are enjoying what they learn are going to show up for class. They are going to be prepared, they will have good self-esteem, and they will be building on what they have learned and apply it. Students are far less likely to slip through the cracks with the common core curriculum in place.

There will still be IEPs (Individual Education Plans) for students that need them. They will be evaluated along with the common core curriculum to see what can be done realistically to help the child to learn and to get to the point where they should be. The process may not be easy but it can be done with the encouragement of educators and parents.

Parent Feedback

Happy kids at school make for happy parents too. They want to know their child is safe and that they are in a good learning environment. They love seeing their child come home excited about what they learned and feeling good about the new concepts. They also enjoy the meetings with parents that show their child is meeting the standards in place.

Parents also like to be able to see that if their child is behind, what exactly they are behind on. The standards in place give the teachers and parent a checklist to go through and to evaluate. If a child is behind on something, then the parent and the teacher can come up with a plan of action to get resources or one on one help in place.

Everyone gets a Voice

The fact that educators, admin, parents, and even students all play a role in creating the common core curriculum means that it can be evaluated on many different levels. There is clear communication in place about what is expected to be taught, what is expected to be learned, and the standardized tests that will conform what was learned.

Everyone knows exactly what needs to be taught, how to help kids learn, and parents know what they should be studying in a given grade. It takes the guesswork out of the equation and makes education a collaborated effort with everyone involved on the same page. Anyone who has had an issue with a public school about curriculum or the difference between what is taught and standardized testing will really appreciate this.

Focusing on the core concepts, understanding, and processes involved needs to start in the very early grades of students. This helps them to build on it and to grow from it. Students need the encouragement and they shouldn't be set up to fail. Graduating from high school aren't enough Students need to graduate with real world stills that they can count on.

It is important to point out that the common core curriculum will continue to be updated over time. As the needs of students for their role in society in the workforce and for college change, so will what is being taught. As the information on assessments gives information, that data will be used to determine what is working well and if there are any weak spots.

This will prevent the common core from becoming obsolete or from not providing students with what they really need from the public school learning environment over time. In fact, it will only continue to get better with time and that is very encouraging to admin, educators, students, and parents.

Alternative Ways to Learn

As adults, we know that all of us don't learn the same way. Some of us are visual learners at work. We have to be shown how to do something versus told. Others can just look at a project or a method that is there and figure out a better way to get it done.

Well, the same is true of children. If math, English, or other subjects are only taught one way, then they may be limited and frustrated. If they need an alternative way to learn, it is easy enough for educators to do so with the common core program. This is going to help ensure fewer kids get behind when it comes to learning the standards for their given grade.

Adjusting teaching methods in order to accommodate the students is very important. Some teachers are very stuck in their methods, regardless of if they are still effective or not. Those methods may have been a huge success a decade ago, but times have changed. Getting teachers up to speed and willing to try some new options is very important.

There may be some training sessions and workshops for teachers to attend that feel they are limited in what they teach. Sometimes, it takes admin evaluating their teaching methods to make such recommendations. Remember, doing so isn't to punish a teacher or to tell them what they are doing is wrong. It is to reach out broadly to students and to make sure everyone is able to benefit from the learning environment.

CHAPTER 9- CONCLUSION

The fact that the common core curriculum has been accepted by all but 5 states is very encouraging. This isn't a program that was federally created, it was the collaborated effort of admin, educators, parents, students, and concerned citizens. Creating a program that allows all students in public schools an equal opportunity to learn makes sense.

The standards for each subject in each grade level are very high. Parents don't have to worry that their children aren't going to have the skills and knowledge that they need. In fact, the very opposite is true. So many parents feel that they have let their children down because they can't afford private school. They worry that the public school they are in isn't offering them enough.

For children that move, the core curriculum means that they can jump right back into their education in the new location and not be too far behind or ahead. The same standards will be used across the board for each grade so it doesn't matter where they attend.

Common Core Mathematics

Measuring the success with the common core gets easier too because the same standard testing can be used for each grade level. This will help to justify the program and to verify students are being taught what they should. Since the standards are designed to ensure students graduate with skills they can use in the world for work and college, it is very comforting and reassuring.

Too many children have been falling through the cracks in public school systems. The common core is designed to prevent that from happening. Educators can still create their lesson plans and teach the way they desire. They just have to make sure they are covering particular standards for each subject.

The standards are very clear and they are consistent. They will change over time too as the expectations in society and in colleges continue to increase. The foundation of what the children learn in each grade will be built upon for the next higher level of education.

With the common core, states that take part in it will be able to share information. They will be able to ensure students get the learning environment they need to be successful in the future. The two biggest areas of changes will be to math and English. Consistently exposing children to materials that are on track and that they can apply is important.

Parents can have a bigger role in the education system and in the learning of their children when the common core is in place. While not all may be initially sold on the idea, the more they learn about it the better they like it. Parents are encouraged to spend time in the classroom and to be involved in what their kids are really taking away from the classroom that they can apply in real life situations.

As the needs of the students move forward, the common core will also be moving forward. As a result, students won't outgrow it and it won't stop helping them to get the education they need to be successful later in life either.

Common core has become increasingly popular because it is no secret there are problems within the public school systems. There is a lack of equality across the board in terms of what students learn. With this new concept, everyone has the same opportunities.

Teachers are embracing it as they really do want to see students succeed. They are tired of taking the blame when there are low test scores. They have argued for quite some time that they don't have the same resources as other educators and public schools. Finally, people are listening.

Parents enjoy common core because they want to see their children grow up and to be self-sufficient in society. Graduating high school and lacking skills for a job, for a college education to be possible, and to have a happy life isn't what they want to see occur for their children.

With common core comes huge changes in the classroom for many students, their teachers, and of course parents. However, it is also going to be changes that are positive and that get results. Many will tell you that this is the way that public education should have been from the very start.

The evaluations in house and with standardized test scores are going to really hold lots of weight when it comes to seeing the true value of the common core curriculum. Some are impatient to see those results, but they will materialize in time. In a poll taken in early 2013, 80% of the parents support the common core curriculum.

Common Core Mathematics

Remember, none of the states that agreed to implement the common core had to lower their standards. If anything, there are states out there that had to increase theirs. It means no child gets left behind or is a victim of a poor education due to where they happen to live. The benefits from the common care are quite encouraging when you look at the big picture!

About The Author

Julie Thompson is well-known educator and a mother of two. Julie Thompson is very keen on the development of children in an institution particularly in public schools. Julie believes that the youth are the hope for the future. This can only be achieved if there is proper guidance and proper education.

Julie Thompson and her family lives in Sacramento with her pet dog.

www.ingramcontent.com/pod-product-compliance
Lightning Source LLC
Chambersburg PA
CBHW051259170526
45165CB00004B/1783